The ether of Albert Einstein

Robert Jobard

The ether of Albert Einstein

What Albert Einstein said that has been forgotten

To my mother,

and to Einstein.

Éther or not éther,
that is the question

Introduction

Albert Einstein's ether, a title that surprised us, as we were amazed to learn that Albert Einstein had reinstated ether after he had suppressed it in 1905. We were also surprised when we had the opportunity to read the French translation of his 1905 paper entitled "On the electrodynamics of moving bodies," which is the founding paper of the special theory of relativity.

The surprise was to discover how Einstein himself criticized his two founding postulates concerning the relativity of motion applied to electromagnetic phenomena. The theory is misnamed because if it is true that the 1905 paper is incomplete, it is the starting point of general relativity and what we call general relativity should be called the general theory of gravitation.

We invite you to examine the observations and remarks made by Albert Einstein, who is the best qualified to speak on this subject. These observations are important, and, yet, articles, books, and videos popularizing relativity never mention them.

Before going any further, there is already an astonishing misunderstanding about the theory of relativity. Many people associate it with the famous equation:

$$E = mc^2$$

This equation does not express anything relative; it is independent of the motion of the reference frame in which matter is transformed into energy or energy is transformed into matter.

The notion of relativity was discovered by Galileo. It concerns motion. Albert Einstein's contribution is to have extended the notion to electromagnetic phenomena, which had been discovered during the 19th century.

The equation $E = mc^2$ is associated with the atomic bomb and the horrors of Hiroshima and Nagasaki. Einstein demonstrated the equation but never worked on the bomb. His fabulous theory of general relativity, which goes beyond relativity to formulate the functioning of celestial bodies in space, made him famous, and this fame has contributed to this association.

Einstein's precursors

Aristotle

Aristotle is one of the giants of philosophy. His teacher at the Academy, Plato, had nicknamed him "the intelligence." Aristotle created a school, the Lyceum, just outside Athens. He was also the tutor of Alexander the Great, who became famous for his conquests and love of knowledge.

The five elements

Aristotle's philosophy separates the world into two parts. At the bottom are the four elements of Empedocles: earth, water, fire, and air ranging from the heaviest to the lightest, which he calls the world of imperfection where everything degrades and dies, and which extends to the Moon. Beyond these four elements, the world is perfect and consists of the ether, the fifth element, the fifth essence, which is quintessence, perfection, and unchangeability. It is the domain of the gods and celestial bodies and fills the whole sky.

Earth: cold, dry, and heavy.
Water: cold and humid.
Air: hot, humid, and light.
Fire: hot and lighter than air
Ether: a divine substance located beyond the Earth's atmosphere.

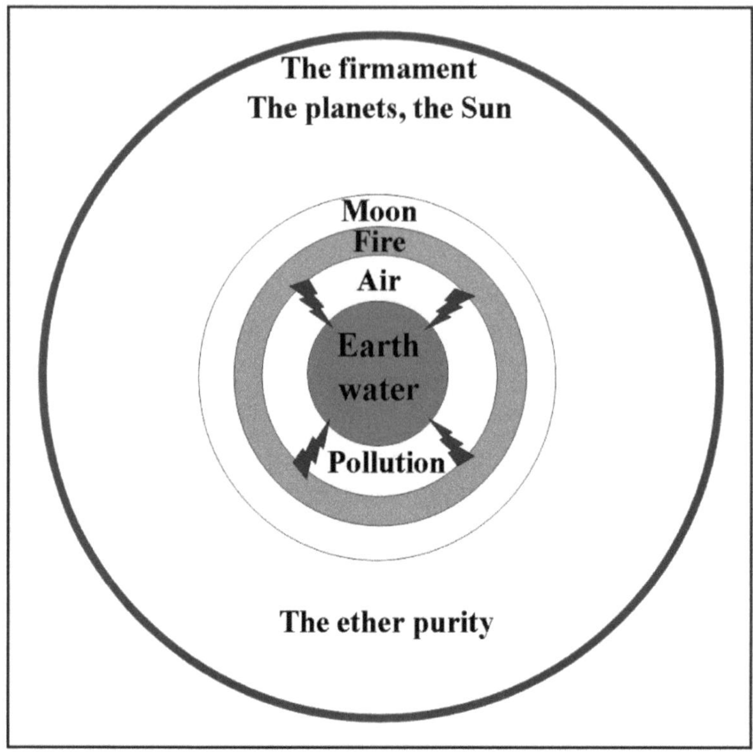

Figure 1. Aristotle's cosmology

Immobility of the Earth

This immobility is undoubtedly Aristotle's greatest error. He asserts that the Earth stands motionless at the center of the world.

He offers an experimental proof:

> We see that if a stone were suspended above some table, it would fall down in a straight line; and as long as its motion is downward it will fall along the same straight line. Now, if the table were not moved, the stone would fall to the [very] same place where it fell before[1]

In other words, heavy projectiles sent vertically upwards fall back to the place from which they were launched. If the Earth rotated from west to east on a daily rotation, the projectiles would have to fall back to the west of the initial point because during their air travel the ground would have moved to the east.

Rather than a stone suspended above some table, we can watch an apple fall.

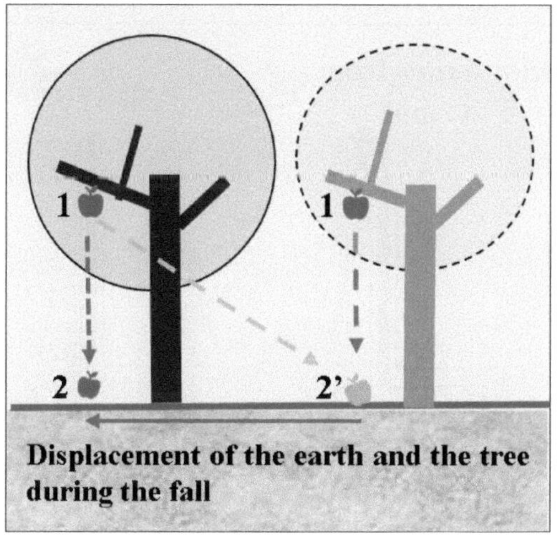

Figure 2. Moving bodies in free fall by Aristotle

The Earth is stationary, otherwise an apple falling from a tree would not fall vertically from 1 to 2 but diagonally from 1 to 2', the position of the apple at the beginning of the fall. The tree would move forward in motion with the Earth.

Giordano Bruno

Based on the work of Nicolaus Copernicus and Nicolaus de Cues, Giordano Bruno developed the theory of heliocentrism and show, in a philosophical way, the relevance of an infinite universe that has no center and is populated by an innumerable quantity of stars and worlds similar to ours.

He observed that a body falling from the top of a ship's mast accompanies the ship during its fall as the ship is moving. This observation negates Aristotle's argument that the Earth is immobile. Galileo explained that this is due to the inertia of bodies, which means that without friction their motion continues indefinitely.

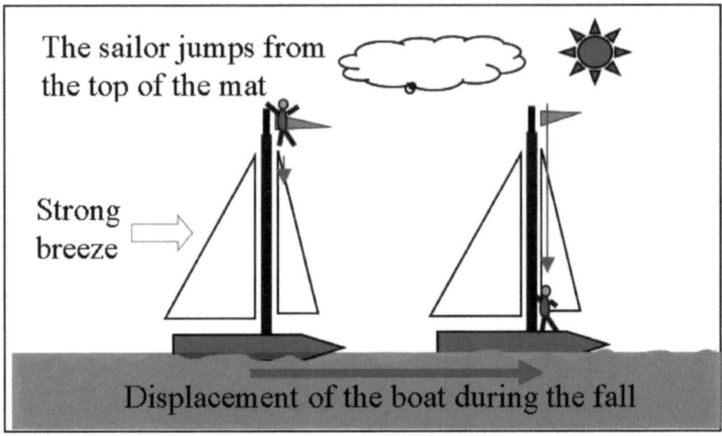

Figure 3. Moving bodies in free fall by Giordano Bruno

Galileo

Mathematician, physicist, astronomer, engineer, Galileo, along with Descartes, is one of the founders of modern mechanics.[2]

The relativity of motion

Galileo established the laws of inertia and the relativity of inertial motion in a straight line. Inertia means that motion in the absence of opposing forces or friction goes on indefinitely. Absolute rest does not exist. It can only be detected by a reference frame located in another reference frame. It is impossible to know which of these reference frames and any others are stationary or not. The motions are relative to each other.

Additivity of speeds

Galileo discovers the additivity of speeds, which he explains very simply by watching the passenger of a boat walk from the back to the front of a moving boat. Seen from the bank, the speeds add up. If the passenger goes backward at the same speed as the boat, seen from the bank, he will appear motionless throughout his journey. The speeds always add up, the speed of the passenger in the boat with respect to the reference frame of the boat is then negative.

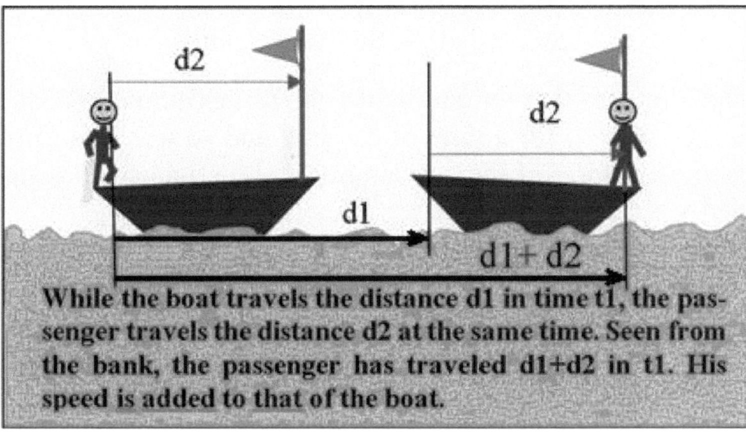

While the boat travels the distance d1 in time t1, the passenger travels the distance d2 at the same time. Seen from the bank, the passenger has traveled d1+d2 in t1. His speed is added to that of the boat.

Figure 4. additivity of velocities by Galileo

The fall of bodies

Galileo studied the fall of bodies. By experimenting with rolling balls of different weights on inclined planes, he discovered that all bodies move in the same way in a uniformly accelerated motion regardless of their weight, and he explained the phenomenon in the following way: the heavier the bodies are, the more difficult they are to set in motion, which exactly counterbalances the fact that the heavier they are, the more forcefully they are attracted by

Earth's gravity. Their resistance to changes in motion corresponds to their inertia. In fact, these explanations were clarified by Descartes and then by Newton, but it was Galileo who initiated this notion.

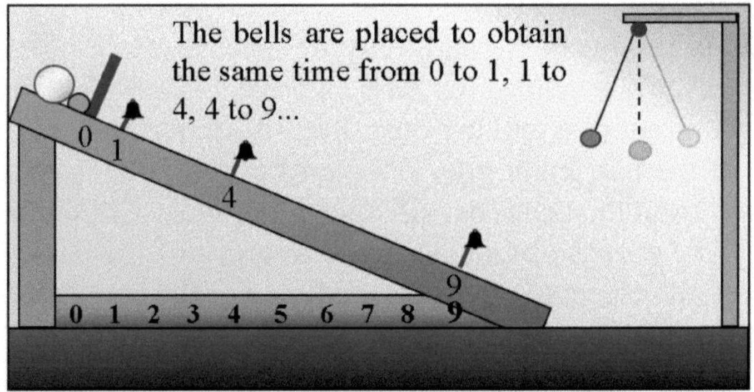

Figure 5. Fall of Bodies by Galileo

Galileo notes that for a time of 1 the distance traveled is 1, for a time of 2 it is 4, for a time of 3 it is 9 and so on. This formula allows the equation of the acceleration to be established: distance traveled = acceleration x time squared.

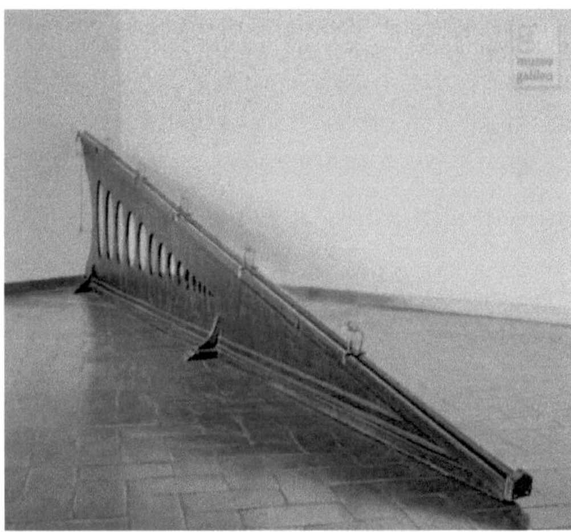

Figure 6. Museo Galileo - Istituto e Museo di Storia della Scienza
https://boowiki.info/art/simples-machines/pente.html

René Descartes

Descartes rejects the theory of the vacuum because it is not possible for nothingness to possess extension. Thus, according to Descartes, if a vase is empty of water, it is full of air, and if it were empty of any substance, its walls would touch.

The mechanism of vortices

As far as the movement of the planets is concerned, Descartes excludes an action at a distance from the Sun, as this idea had no rational basis at that time. He opposes an empty space and fills it with ether swirling around the Sun and causing the movement of the planets.

In Descartes' theory, the ether drags the Earth and moves with it. Locally there is no motion between the Earth and the ether, and the ether is no longer an absolute reference frame.

https://fr.wikipedia.org/wiki/Théories_scientifiques_de_Descartes

Isaac Newton

Famous physicist, philosopher, and mathematician; at the University of Cambridge, Newton broke down white light with a prism and invented the reflecting telescope by replacing the telescope lens with a reflecting concave mirror. Around 1700, he published the "Mathematical Principles of Natural Philosophy," which establishes the law of universal gravitation: celestial bodies are attracted to each other by a force of intensity proportional to their masses and inversely proportional to the square of the distance between them.

$$F = G\ Mm'/d^2$$

Christiaan HUYGENS

Dutch mathematician, astronomer, and physicist. Huygens was an elected "fellow" of the Royal Society of London and subsequently an eminent member of the Royal Academy of Sciences founded by Colbert in Paris. He also participated in the realization of the Observatory of Paris.

In 1672, he began to study the nature of light and imagined light as a vibration, which allowed him to explain the phenomena of diffraction. Later, Thomas Young and Augustin Fresnel would also make their contributions. The transverse nature of light waves explains all polarization phenomena. Hippolyte Fizeau and, sometime later, Léon Foucault measured the speed of light in water and showed that its speed was lower than in air. Newton's corpuscular theory asserted the opposite. The wave theory triumphed.

Joseph Louis, LAGRANGE

Astronomer and mathematician. Lagrange published a treatise on "Analytical Mechanics," gathering the results of his calculations on the stability of the solar system and showing that the inequalities in the motions of the planets are in fact variations with very long periods.

Most of the fundamental equations of physics can be formulated from the principle of least action.

The Lagrange functions and the "Lagrangian," which allows the equations of the motion of the dynamic variables of a system to be described, are always used.

Physics at the end of the 19th century

James Clerk Maxwell published the equations of electromagnetism in 1865. From his equations, he predicted the existence of waves associated with electromagnetic fields. He calculated the velocity from the dielectric permittivity ε (Coulomb's law) and the magnetic constant μ (Ampere's theorem) of the medium studied. In a vacuum, these data have the index "0." Coulomb's law and Ampere's theorem allow us to determine them by measuring the effects produced.[3] Here is the old equation used by Maxwell:

$$c_0 = \frac{1}{\sqrt{\varepsilon_0 . \mu_0}}$$

And the modern presentation $\varepsilon_0 = 1/c_0^2 \mu_0$

You will note in passing that the so-called vacuum has physical properties: the dielectric permittivity and the magnetic constant, which are not zero. If they were zero, the speed found would be equal to infinity and we could speak of an absolute vacuum.

The calculation gives a result close to 300,000 kilometers per second, which is approximately the same value as that of light. Maxwell deduced that light is itself an electromagnetic wave. Waves need a medium to propagate. This support will be called ether to honor Aristotle but will be different from the fifth element. It must be present everywhere, without any friction and have an elasticity associated with a great rigidity so that the electromagnetic waves propagate at the fantastic speed of 300,000 km/s.

The knowledge of the universe in the 19th century was still very limited. We did not know the galaxies. We imagined all the fixed stars as well as the Sun, which is our star. With the Earth not being the center of the Universe, one associated the reference frame of the ether with that of the fixed stars, the reference frame of the universe.

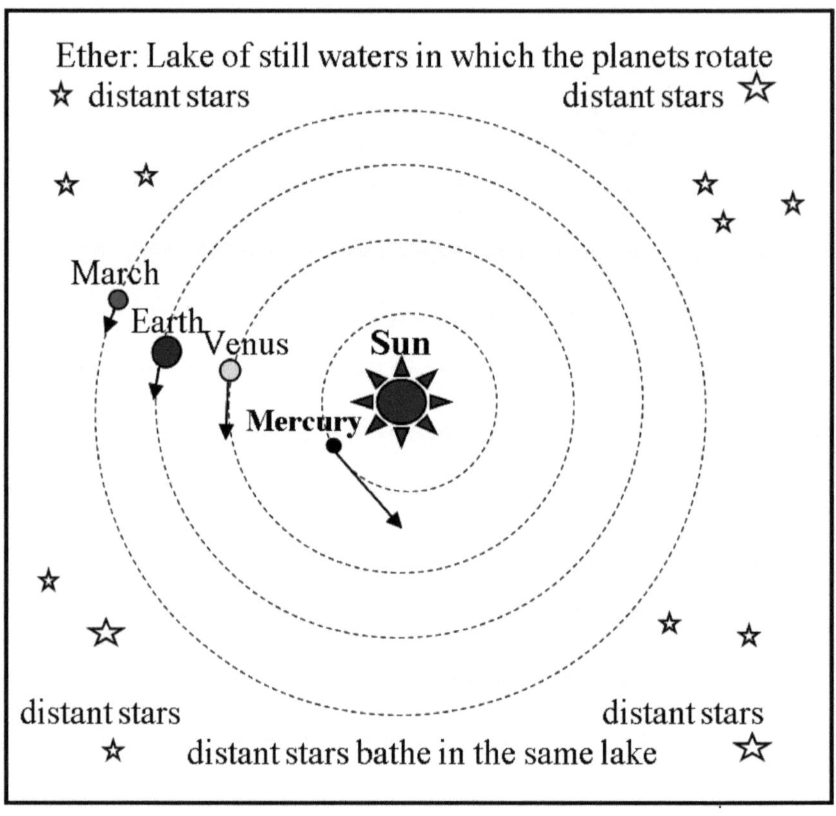

Figure 7. The vision of ether in the 19th century

Michelson and Morley experiment

Michelson proposes to demonstrate the movement of the Earth in the immobile ether in relation to the Sun and the stars. He imagined building an interferometer that would show the difference in the speed of light between a ray of light moving parallel to the Earth's trajectory and affected by the speed of the Earth and another moving perpendicularly. With Morley, he built a very rigid apparatus protected from vibrations.

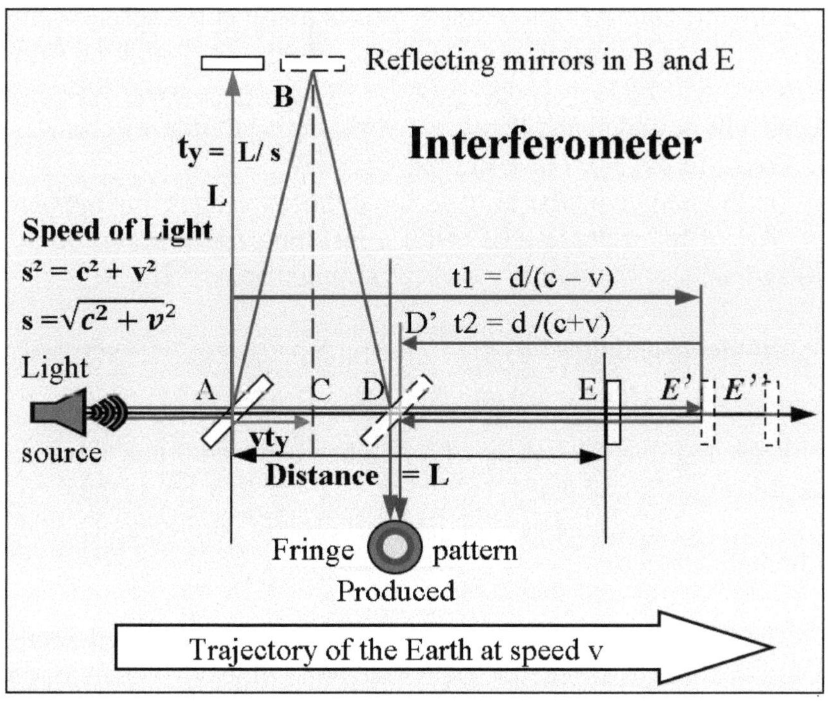

Figure 8. Michelson–Morley experiment

This experiment is very well known; it is the most famous failed experiment in physics. The diagram on the previous page shows the three positions of the semi-reflecting mirror inclined at 45°, in A when the light ray arrives, in C when the deflected ray bounces on the mirror in B, and in D when it returns. The undeflected ray is sent back to the screen in D' with a slight delay.

A light ray is sent to a semi-reflective mirror, which deflects one half of the ray towards a mirror located at 90° of the trajectory and lets the other pass, which reaches a mirror located at the same distance. On the way back, the semi-reflecting mirror again deflects the rays, which will display the interferences they produce on a screen and whose distance allows the speed difference between the two rays to be calculated.

Forward path AE: the point E moves away at the speed v, the distance d to be covered will be at the speed $c - v$, and the time necessary will be $t1 = L / (c - v)$. On the return journey E'D, the speeds will be added and $t2 = L / (c + v)$; the total time will be

$tx = L/ (c - v) + L / (c + v) = 2L\ c/(c^2 - v^2)$.

On the perpendicular branch, the light ray follows the hypotenuse AB of the triangle ABC with a resulting speed according to Pythagoras: $s^2 = c^2 - v^2$ or $s = \sqrt{(c^2 - v^2)}$ round trip $ty = 2L / s$

$ty = 2L/\sqrt{(c^2 - v^2)}$

Divide up and down by c^2:

$tx = 2\ L/c\ 1/(1 - v^2/c^2)$ and $ty = 2\ L/c\ 1/\sqrt{(1 - v^2/c^2)}$

Knowing that the square of a number between 0 and 1 is less than this number and that $1 - v^2/c^2$ is < 1, $1 - v^2/c^2$ is lower than $\sqrt{(1 - v^2/c^2)}$ and therefore $ty < tx$. The travel time on the perpendicular branch is shorter.

The displacement of 30 km/s of the Earth could not be demonstrated by the measurements. Deviations of 8 km/s were mentioned which were attributed to measurement errors It is now known that the Sun and stars are in motion in the universe where they gather in galaxies and that these galaxies move. The experiment of Michelson and Morley was doomed to failure, ether or not.[4]

However, at the time it was a great surprise. There was no question about the ether accompanying the Earth. The Earth could not be the center of the universe. It revolved around the Sun with the planets.

In fact, the same error was repeated, not with the Earth but with the Sun, which could not be the center of the universe either.

Lorentz's theory of the ether

Lorentz sought an explanation for the absence of motion of the Earth in the ether. He introduced a strict separation between matter, electrons, atoms and their nuclei, which were not yet known at the time, and ether.

He describes the state of ether from the electric field E and the magnetic field H, which are caused by the states of excitation or vibration of the electron charges. This description represents an abstract electromagnetic ether that replaces the previous mechanical models of ether.[5]

In 1889, Oliver Heaviside had shown through Maxwell's equations that the electric field surrounding a spherical distribution of charge should cease to have spherical symmetry once the charge is in motion relative to the luminiferous ether. FitzGerald then conjectured that Heaviside's distortion result might be applied to a theory of intermolecular forces. Some months later, FitzGerald published the supposition that bodies in motion are being contracted.

$$\sqrt{1 - v^2/c^2} \quad \text{(Heaviside ellipsoid).}$$

In 1895, Lorentz proposed various possibilities concerning the contraction of the interferometer to explain the phenomenon:

1) The contraction is in the direction of motion.
2) A dilatation occurs in the directions perpendicular to the motion.

3) The interferometer contracts in the direction of the motion and expands in the perpendicular directions, with the two effects adding up to give the desired quantity.

Finally, the Lorentz contraction occurs in the direction of motion and without perpendicular dilation, which is expressed with the factor $\sqrt{(1- v^2/c^2)}$:

$$L_0 = L \sqrt{1 - v^2/c^2}$$

Time is also affected in the moving reference frame. The notion of "local" time is expressed with the equation

$$t' = t - vx/c^2$$

This equation leads us to say a word about Woldemar Voigt, who was the first to propose this "local" time equation. In 1886, Voigt studied the problem of landmarks that leave the speed of light invariant. He was the first to form equations related to the Lorentz transformation and demonstrated the invariance of the wave equation in this transformation.

Voigt corresponded with Lorentz in 1887 and 1888 on the Michelson-Morley experiment. Joseph Larmor was familiar with Voigt's transformation. Voigt's pioneering work was cited in 1903 in the Annalen der Physik and Lorentz, in his subsequent book "Theory of Electrons" cited Voigt as a precursor.[6]

Poincaré finalized the calculations, demonstrated the group property of the Lorentz transformations, and formulated the relativistic velocity composition theorem.

The expression $x^2 + y^2 + z^2 - c^2t^2$ is an invariant and $ct\sqrt{(-1)}$ is the fourth coordinate of a space-time that will be taken up and completed by Minkowski, who will exploit the use of the four-ectors composed of the three-dimensional vectors linked to the x, y, and z axes and the time vector t. The graphical representations are limited to three dimensions. Graphs that include time are limited to one dimension of space or two in a plane perpendicular to the time axis presented in perspective.

The transformations are a consequence of the principle of least action and have group properties, hence, he gave them the name the Lorentz group. Henri Poincaré wrote a paper in July 1905, known as the "Palermo paper," which was not published until 1906.[7]

That is it. We have set the scene. 1905 is the wonderful year of Albert Einstein. He published four articles. The first addresses the photoelectric effect, the second with Brownian motion. The third article, "On the electrodynamics of moving bodies," describes the theory of special relativity, which we will examine now. The fourth describes the mass-energy equivalence, that is, E = mc².

You will notice that of these four articles, only the third one, "On the electrodynamics of moving bodies, " has been violently criticized. We can deduce that it was not Albert Einstein who was personally targeted but this article. We will see why.

On the electrodynamics of moving bodies

Here we have gathered the information useful for a better understanding of the relativity of motion. We are prepared to examine the founding article of relativity published by Albert Einstein in 1905 and highlight the criticisms he made of his theory. Even before applying Galileo's principle of relativity to electromagnetic phenomena, Einstein states that it is contradicted by Maxwell's equations, which for the phenomenon of induction distinguish the case in which the magnet moves from the case in which the coil moves:

> It is known that Maxwell's electrodynamics—as usually understood at the present time—when applied to moving bodies, leads to asymmetries which do not appear to be inherent in the phenomena. Take, for example, the reciprocal electrodynamic action of a magnet and a conductor. The observable phenomenon here depends only on the relative motion of the conductor and the magnet, whereas the customary view draws a sharp distinction between the two cases in which either the one or the other of these bodies is in motion.[8]

Let us turn to the two fundamental postulates of special relativity:

> We will raise this conjecture (the purport of which will hereafter be called the "Principle of Relativity") to the status of a postulate, and also introduce another postulate, which is only apparently irreconcilable with the former, namely, that light is always propagated in empty space with a definite veloci-

ty c which is independent of the state of motion of the emitting body.[9]

Einstein based his entire theory on these two postulates. After admitting that the first postulate contradicts Maxwell's equations, he admits that the second postulate he presents is "at first sight [...] incompatible with the first."[9]

He adds that the ether is superfluous, hence, he will not use it in his calculations. His theory does not need the ether, which is perfectly correct; he does not have to take into account the deviation of the induction equations. However, special relativity uses waves that move in a way that is not specified.

Let us return to the argument "light propagates in empty space, at a speed V independent of the state of motion of the emitting body." This is a characteristic of waves. The speed of a wave does not depend on the speed of the object that produced it but on the physical characteristics of the medium in which it propagates. Waves on the surface of water propagate at the same speed whether the pebble that produced them was thrown very hard or just dropped and regardless of the size of the pebble. The pebble's size and speed will influence the height of the waves, not their speed of travel.

The same is true for sound waves in the air. The speed of sound depends on the temperature of the air. At 15°C at sea level, the speed of sound is approximately 340 m/s or 1,224 km/h. At 10,000 meters, the cruising altitude of airplanes, the temperature is lower and the speed of sound is approximately 300 m/s or 1,080 km/h. In water, the sound propagates at 1,500 m/s and in iron at approximately 5,960 m/s.

You will note that this does not make the speed of sound a universal constant. To its speed must be added the speed of its supporting medium. If a strong wind blows eastward at 80 km/h, the air will move faster eastward by 80km/h and slower westward by 80 km/h. Galileo's speed additivity works.

However, the speed of sound is not additive with the speed of the object that produces it. An airplane can exceed its own sound by flying faster. It is then said that we have exceeded the sound barrier. The notion of a barrier comes from the shock wave that is formed at the front of the plane by the captured sound, which leads to a strong increase in pressure.

Figure 9. **https://pixabay.com/fr/mur-du-son-bris%C3%A9es-marine-jet-964218/**

Let us return to the article on special relativity. The interpretation of photonic phenomena, intelligently published by Albert Einstein just before the article on special relativity, with his massless particle of light carrying energy, allowed him these liberties.[10] This massless particle was called a photon. Since then, physicists have renamed this photon a "wave packet."

§ 1 Definition of simultaneity

We have to take into account that all our judgments in which time plays a part are always judgments of simultaneous events. If, for instance, I say, "That train arrives here at 7 o'clock," I mean something like this: "The pointing of the small hand of my watch to 7 and the arrival of the train are simultaneous events.[11]

This is an excellent definition of simultaneity. When everything happens in the same place, everything is simple. If the measurements are made at points that are distant from each other and if the points are stationary with respect to each other, they are all in the same reference frame and the distances between them remain constant at all times, whatever the elapsed time. Clocks are useless.

However, this simultaneity is not the case when the points are in motion relative to each other or to the observer who is making the measurements. Let us take the following example: I am standing at the beginning of the platform to note the time the train enters the station; if I am placed at the other end of the platform, there will be a certain amount of time between the moment the train enters the station and the moment I see it, that is, the time it takes for the light of the event to reach me. Over such a small distance, the gap is naturally almost indiscernible.

Einstein proposes a method to check the synchronization. Let two observers equipped with a clock stand at point A and point B:

We have not defined a common "time" for A and B, for the latter cannot be defined at all unless we establish by definition that the "time" required by light to travel from A to B equals the "time" it requires to travel from B to A.

Let a ray of light start at the "A time" from A towards B, let it at the "B time" be reflected at B in the direction of A, and arrive again at A at the "A time."

In accordance with the definition, the two clocks synchronize if

$$t_B - t_A = t_{A'} - t_B$$

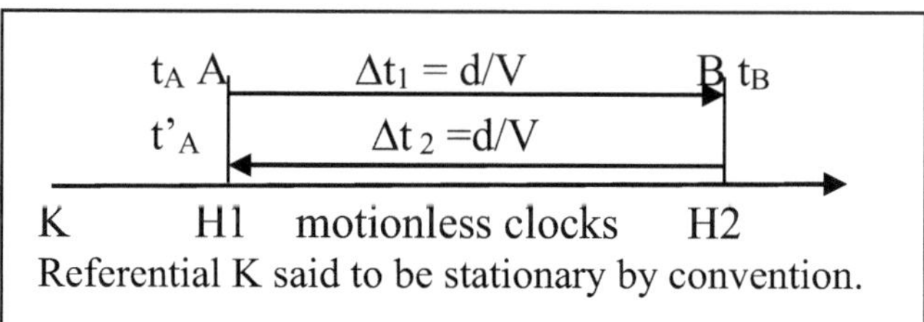

Figure 10. Synchronization of clocks

Einstein observes,

> [...] In agreement with experience we further assume the quantity
>
> $$\frac{2\overline{AB}}{t'_A - t_A} = V,$$
>
> to be a universal constant—the velocity of light in empty space.
> It is essential to have time defined by means of stationary clocks in the stationary system, and the time now defined being appropriate to the stationary system we call it "the time of the stationary system."[12]

§2 On the relativity of lengths and times

The laws by which the states of physical systems undergo change are not affected, whether these changes of state be referred to the one or the other of two systems of co-ordinates in uniform translatory motion.

Any ray of light moves in the "stationary" system of co-ordinates with the determined velocity c, whether the ray be emitted by a stationary or by a moving body. Hence [13]

$$\text{velocity} = \frac{\text{light path}}{\text{time interval}}$$

Light is independent of the speed of the object that produces it. This property is a characteristic of waves; their speed is determined by the characteristics of their medium. An airplane can catch up with and exceed the sound it produces. This does not make the speed of sound an absolute value that is the same everywhere.

Consider a rigid rod at rest; it has the length l when measured by a ruler at rest. What is the length of the rod in motion?

Here we come to the important point, the way in which Einstein observes a frame of reference k in motion from another frame of reference, K, which is labeled stationary by convention.

(a) The observer moves together with the given measuring-rod and the rod to be measured and measures the length of the rod directly by superposing the measuring-rod, in just the same way as if all three were at rest.

(b) By means of stationary clocks set up in the stationary system and synchronizing in accordance with § 1, the observer ascertains at what points of the stationary system the two ends of the rod to be measured are located at a definite time. The distance between these two points, measured by the measuring-rod already employed, which in this case is at rest, is also a length which may be designated "the length of the rod.[13]

We measure the ruler AB of length L. The observer of k and the observer of K both find in their reference frame the same length L. It is stationary with respect to them. The observer of K then decides to measure the length r_{AB} of the ruler, which is in the reference frame k moving at the speed v along the x-axis. On the outward journey, point B moves away at speed v

On the way back, point A is approaching at the speed v

Einstein concludes that the clocks of k are not synchronized, seen from the K frame of reference.

Let us note that according to Albert Einstein, in k everything happens as in K, and seen from K the lengths in the direction of the motion contract and the time dilates in small k. This deduction will allow us to go on to the paradox of the twins.

§ 3. Theory of the transformation of co-ordinates

Here Albert Einstein will transfer the dimensions of space and the time scale from the frame k into the frame K. He uses the rigid rulers to measure, and the synchronized clocks seen previously. Consider a system K called stationary with the co-ordinates x, y, z and t for time and a system k with the co-ordinates ξ, ψ, ζ and τ which moves parallel to K along the x-axis at speed v.

> We now imagine space to be measured from the stationary system K by means of the stationary measuring-rod, and also from the moving system *k* by means of the measuring-rod moving with it; and that we thus obtain the co-ordinates x, y, z, and ξ, η, ζ, respectively. Further, let the time **t** of the stationary system be determined for all points thereof at which there are clocks by means of light signals in the manner indicated in §1; similarly let the time τ of the moving system be determined for all points of the moving system at

which there are clocks at rest relatively to that system by applying the method, given in §1, of light signals between the points at which the latter clocks are located

[...]

If we place *x'=x-vt*, it is clear that a point at rest in the system *k* must have a system of values x', y, z, independent of time. We first define τ as a function of x', y, z, and t. To do this we have to express in equations that τ is nothing else than the summary of the data of clocks at rest in system *k*, which have been synchronized according to the rule given in § 1.

From the origin of system *k* let a ray be emitted at the time τ_0 along the X-axis to *x'*, and at the time τ_1 be reflected thence to the origin of the co-ordinates, arriving there at the time τ_2; we then must have

$$\frac{1}{2}(\tau_0 + \tau_2) = \tau_1.$$

or, by inserting the arguments of the function τ and applying the principle of the constancy of the velocity of light in the stationary system:

$$\frac{1}{2}\left[\tau(0,0,0,t) + \tau\left(0,0,0,t + \frac{x'}{c-v} + \frac{x'}{c+v}\right)\right] = \tau\left(x',0,0,t + \frac{x'}{c-v}\right).$$

Or

$$\frac{\partial\tau}{\partial x'} + \frac{v}{V^2 - v^2}\frac{\partial\tau}{\partial t} = 0. \quad [14]$$

In the above we have preferred to reproduce the original English translation than to make our own comments. We would not want to risk mistranslating the thought of Albert Einstein.

Below we present a graph summarizing this delicate passage.

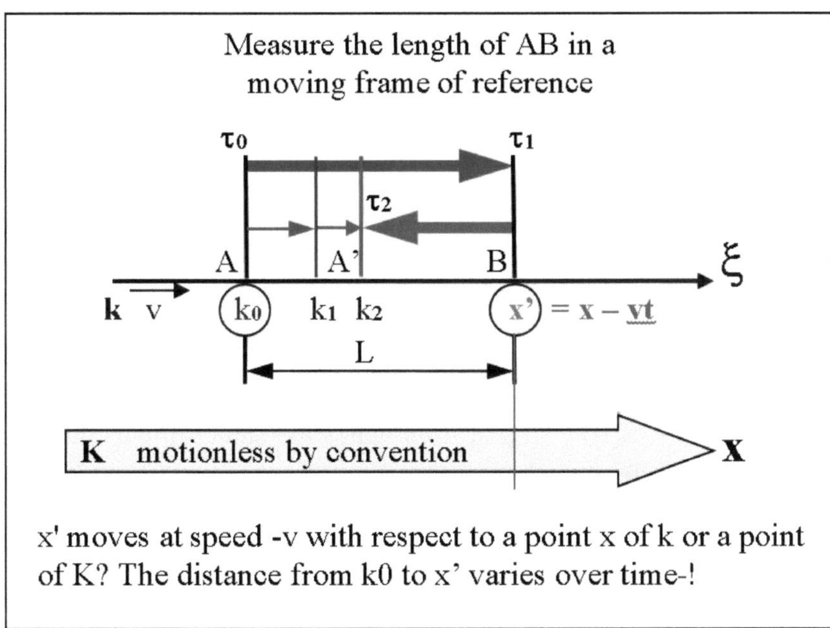

Figure 11. Measurement of the length of a rod in a reference
frame in motion with respect to the reference frame of the observer
This graph should clarify the situation.

For the remainder of the theory you can refer to Einstein's article. It is about the mathematical developments that led to finding the Lorentz-Poincaré transformations.

Here we are, we have taken the time to deconstruct Einstein's reasoning. Usually, explanations are limited to the Michelson-Morley experiment. Einstein does not try to explain the failure of the experiment through a contraction of matter but by suppressing the ether and especially its absolute reference frame in the universe.

He did not succeed in freeing himself from all absolutes. He made the speed of light an absolute and recognized that this contradict his first postulate which, by applying relativity to electromagnetic phenomena in motion, forbade any absolute.

We note that the fact that the time taken to go from clock H1 to clock H2 by an electromagnetic wave must be the same on the outward journey as on the return journey does not prove that these clocks are synchronized, but that they can be synchronized because between them they are immobile. When the times are different, the clocks are in motion with respect to each other.

We could try to synchronize them by adding the time t1 given by the signal to arrive so that it indicates the same time as the first clock, whose time has also progressed by t1. On the way back, we could check whether the time of the second clock plus the return time t2 is also synchronized with the original clock. Let us not forget that what we call "local time" is the time as we see it in the moving frame of reference. It is the local time of this frame of reference for an observer outside this frame of reference. For an observer of the moving frame of reference, time and lengths are unchanged, and this is the basis of the notion of the relativity of motion.

For an object that moves away, the time taken by a signal to go from its closest extremity to the one that is farthest away is slower, and, in contrast, the signal returns more quickly. If, conversely, the object is approaching, the opposite occurs. Should we not distinguish the two cases rather than averaging them?

In the case of the Doppler effect, the sound emitted by an approaching object is more acute, and it becomes less sharp when the object moves away.

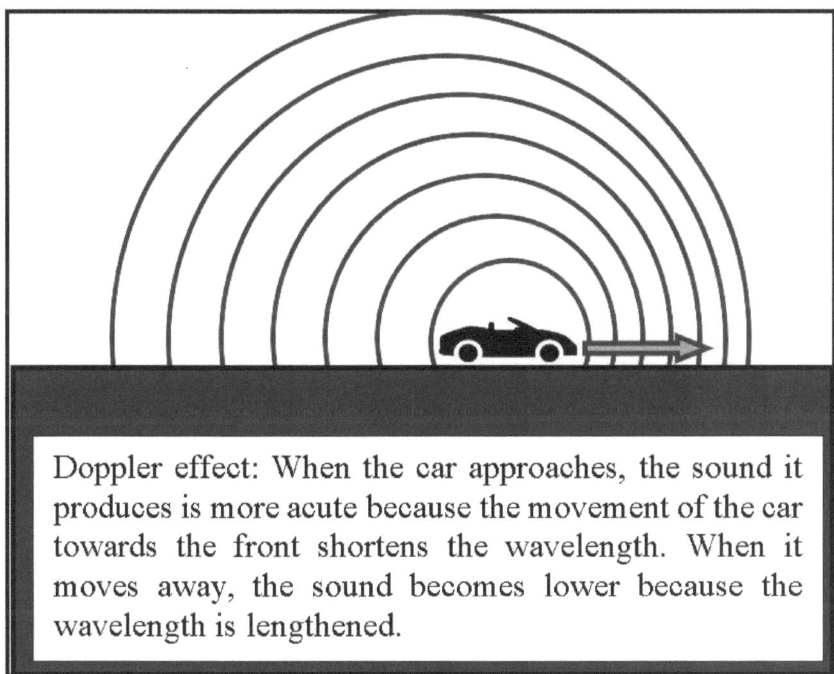

Figure 12. Doppler effect

Well, we have tried to make things as clear as possible. These last remarks are mainly intended to bring us to the paradox of the twins.

The paradox of twins

Paul Langevin's twin paradox, submitted for the approbation of his colleagues at the Bologna Congress in 1911[15], states that twins separate at the age of 20, and one of them leaves for an inter-stellar journey in a rocket moving at a speed close to that of light, such that, seen from the Earth, the dilation of time means that one second for the rocket lasts two seconds on Earth. After 20 years, those on Earth see the rocket making a U-turn and returning to Earth at the same speed to arrive 20 years later, after a journey of 40 years. The twin who stayed on Earth is 60 years old. He welcomes his traveling brother, who is only 40 years old. He has aged by half the number of years because of the time dilation observed by the Earth.

The paradox lies in the fact that the traveler has aged by 40 years in his own frame of reference in motion, since in his frame of reference he was not in motion. It is the Earth that moved away and then returned in the space of 20 years, while the traveler lived 40 years. So it is the traveler who is 60 years old and finds his 40-year-old brother.

A smart guy then points out that each of the twins lived 40 years in their own landmark, so they are both 60 years old.

Faced with these three possible answers, the most prevalent opinion in the literature states that the traveler has aged less, on the grounds that at the departure of the rocket, when it makes a U-turn and at its arrival, it is no longer in an inertial Galilean reference frame but in accelerated reference frames. Such was the opinion of Albert Einstein.

We do not share this idea: The non-inertial reference frames are supposed to be neglected; if there is a difference in age it will be due to the accelerations that affect time in the traveler's reference frame. If we are only interested in the inertial part of the frame of reference, the time flows in the same way as on the Earth. For us, the twins have both lived 40 years in their own frame of reference and are both 60 years old.

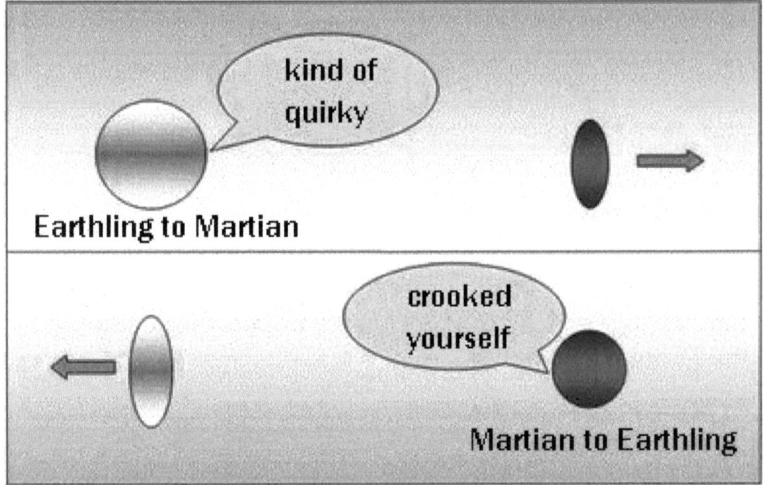

Figure 13. Twin paradox

We have a strong argument to support this solution, that is, what Albert Einstein wrote in his 1905 article in §2 On the relativity of lengths and times:

Let a rigid rod be at rest; it is of length 1 when measured by a ruler at rest. What is the length of the rod in motion?

a) The observer with the ruler moves with the rod to be measured and measures its length by superimposing the ruler on the rod, as if the observer, the ruler, and the rod were at rest.

b) The observer determines at which points of the stationary system the ends of the rod to be measured are located at time t, using the clocks placed in the stationary system (the clocks being synchronized as described in § 1). The distance between these two points, measured by the same measuring-rod when it was at rest, is also a length, which we call the "length of the rod"

Who can say it better?

General relativity

The history of relativity is not finished. General relativity is Albert Einstein's masterpiece. He said he had the "happiest idea of his life"[16] in 1907 when he realized that in free fall one does not feel one's own weight, one does not feel subjected to any force, as if one were floating in a vacuum. He deduced that no physics experiment performed in a frame of reference in free fall in a gravitational field could distinguish this state from that of an inertial frame of reference at rest far from any gravitational field. This makes free fall the equivalent of an inertial reference frame such as those of special relativity. From there, he tries to generalize inertial relativity to relativity in a gravitational field.

At the same time, he imagines an elevator far from any gravitational field. In this inertial reference frame, if the elevator car accelerates towards the ceiling, the notion of up and down appears, and the passenger finds a more or less high weight according to this acceleration. He is attracted by the floor of the elevator car and the objects start to fall down. This analogy allowed Albert Einstein to establish the equivalence between acceleration and gravitation.[17]

This theory calculates the deformation of space as a function of the density of matter-energy present in this space. The celestial bodies in motion follow these deformations because of the principle of least action. This principle means that the trajectory of an object is the one that requires the least effort. This path is called geodesic. A geodesic is a sinuous path that depends on the curvature of space at the location and that respects the principle of least action, which corresponds to free-fall trajectories.[18]

Curved spaces lead to complex equations. In the summer of 1915, the mathematician David Hilbert asked Einstein to come to Göttingen to present general relativity. These lectures allowed Hilbert to understand the problem so well that on July 15, 1915 Einstein wrote to Arnold Sommerfeld,

"In Göttingen I had the great joy to see everything understood and even in detail. I am very enthusiastic about Hilbert. A great man!"[19]

Einstein presented his theory on November 25, 1915 at the Berlin Academy. It turns out that on November 20 or 23, 1915, Hilbert also presented a theory to the *Mathematische Gesellschaft* in Göttingen. The gravitational field equations of these two theories were very similar. Hilbert had added two very ambitious axioms to his theory: a law of physical evolution determined by the world function H and that this world function H should be invariant to parameter transformations.

A controversy developed for a time regarding whether Einstein or Hilbert had formulated the equation first. Here, again, this controversy is irrelevant. Hilbert did not claim authorship of the equation, and it is also likely that without the presentations of Einstein, who had been working on the question for seven years, Hilbert would not have had the opportunity to become interested in it.

Gravitational waves

The principles of special relativity lead to the postulation that the gravitational interaction propagates at the speed of light. In 1905, Albert Einstein had the photon to justify the movement of light. In 1916, failing to find a graviton, he realized that his gravitational waves must move in a space that is not empty.[20]

Under the influence of Lorentz, who developed a theory of ether,[21] and the German physicist Philipp Lenard, Einstein then recognized the possibility of introducing a new concept of ether. The state of this "new ether" would determine the motion of physical objects, whose metric behavior would be described by the gij tensor of his equation (the space curvature tensor). However, he believed that these ideas were not very clear.

Existence of a medium of support for electromagnetic waves

In 1920, the University of Leiden offered Albert Einstein a position, which he accepted. He gave an acceptance speech,[22] in which he presented his vision of the ether and published the speech in 1921. This text constitutes his first major work on the ether. He described Lorentz's work and pointed out that Lorentz stripped the ether of all mechanical properties except one, its immobility. The theory of special relativity removes this last mechanical property of the ether. According to the principle of relativity, the laws of physics – including those of electromagnetism – are identical in all frames of reference in straight and uniform translations with respect to each other. There is therefore no physical reason to distinguish a particular reference frame in which the ether would be immobile, creating an asymmetry that is not justified or detected by any physical experiment. This constraint is difficult to explain; the motion of the ether must adapt to each reference frame. [23]

In 1921, Einstein published an article entitled "The Ether and the Theory of Relativity" sometimes presented under the title "Sidelights on Relativity." We find his conclusion there:

"Recapitulating, we may say that according to the general theory of relativity space is endowed with physical qualities; in this sense, therefore, there exists an ether. According to the general theory of relativity space without ether is unthinkable; for in such space there not only would be no propagation of light, but also no possibility of existence for standards of space and time (measuring-rods and clocks), nor therefore any space-time intervals in the physical sense. But this ether may not be thought of as endowed with the quality characteristic of ponderable media, as consisting of parts which may be tracked through time. The idea of motion may not be applied to it."[24]

Could it be that what Albert Einstein calls space in his theory of general relativity is the ether itself? A space that deforms according to what it contains is not a space in the usual sense of the term. Space represents a more or less large volume within which objects are placed and evolve independently. Space has no physical properties; it is just the description of the volume in which matter and energy can evolve according to their own physical properties.

If there is a deformation, it means that something in the considered space is deformed, and then it is no longer "space" but rather this thing that we are discussing. In his popular science lecture in 2016, the professor of theoretical physics at the Institute for Advanced Scientific Studies, Thibault Damour, compares space to a block of jelly that vibrates.

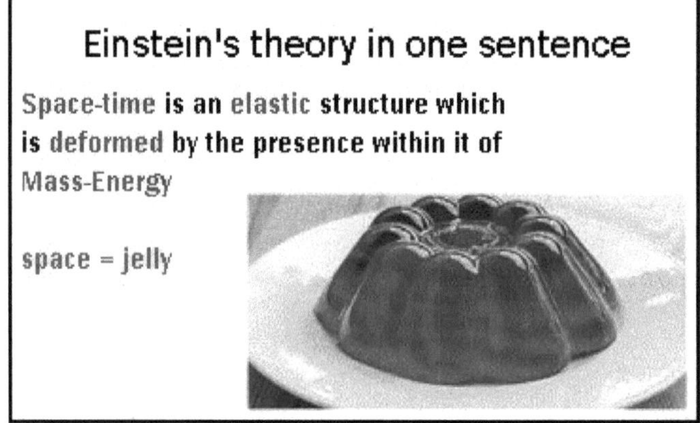

Figure 14. Thibault Damour : Space is a block of jelly[25]

Albert Einstein has been heavily criticized for having suppressed the support of electromagnetic waves. His choice to reinstate this support has, it seems, left physicists indifferent.

There are many indications that space is not empty of everything. Pairs of particles of matter and antimatter emerge from it to disintegrate rapidly and return to this vacuum, which we now know contains a large quantity of energy, the so-called vacuum energy – a beautiful oxymoron!

There is also the pressure exerted by the vacuum on two very close plates, the Casimir effect.

In 1948, Hendrik Casimir predicted, from the quantum fluctuations of the vacuum, that two plates very close to each other attract each other.

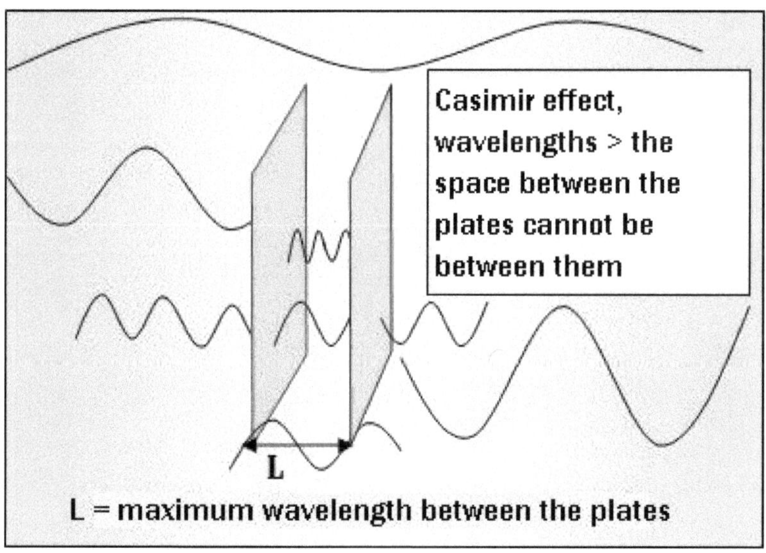

Figure 15. Casimir effect

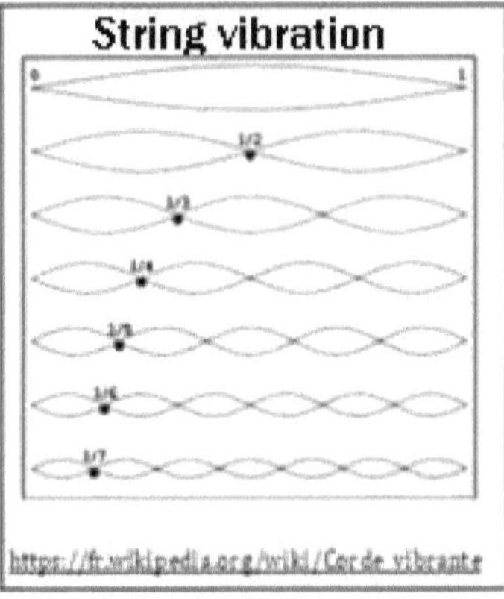

Figure 16. string vibration

This attraction results from the radiation pressure, which is stronger on the outer faces than on the inner faces, where only equal or sub-multiple wavelengths of the distance L can be formed.[25]

New hypothesis for ether

We have now arrived at the goal. Albert Einstein recognizes the existence of an ether, the Lorentz ether, from which he withdraws the immobility. The ether must accoit isitmpany the Earth in its movement but also all the reference frames of the universe.

It is enough to assume that the ether, matter or energy, is subject to the laws of gravitation.

In relativistic language, we will say the ether follows the geodesics of space-time.

So it accompanies the Earth on its geodesic.

For the same reasons, it must accompany Mars, Venus, or any celestial body, star, or galaxy on their own geodesics. As Galileo explained, all bodies fall at the same speed regardless of their mass. The ether accompanies all celestial bodies in their movements.

It would be as simple as that!

When Albert Einstein states that the ether is affected by matter, and vice versa, this is perfectly in line with our hypothesis. Matter acts on the ether through gravitation, and the ether accompanies planetary systems in their whirl around their star, always because of gravitation. The star takes ether with its planetary system in its race around its galaxy. The galaxy, for its part, travels along the great galactic rivers, which are also geodesics and which converge towards great attractors.

The wandering stars

Our hypothesis supposes the presence of a medium filling the whole universe. We have explained the absence of friction by putting it in motion with all the celestial bodies in "free fall." It is normal to ask the question with regard to wandering stars such as comets and asteroids. The objects in motion in the vacuum undergo a heating called the Unruh effect,[26] named after its discoverer. We will assume that the wandering stars undergo the friction of the ether as long as they are not on a stabilized trajectory.

After this presentation, you might say to yourself, "it is too simple; if it were true it would have been discovered a long time ago." This point is true, but it is important to note that all the explanations were formulated by Albert Einstein himself. His answer was in his own theory. Our work was laborious, but in the end it
was simple, we just had to read what he had written. This hypothesis is rightfully his.

We are not proposing any physical properties for Einstein's ether, but we have nevertheless given it back its motion. The question will arise if its existence is verified. To speak about the deformation of space in the course of time is the abstract view of a mathematician, it is very precious. The physicist has the duty of trying to understand what physically deforms and is able to force moving bodies to follow the geodesics of gravitation.

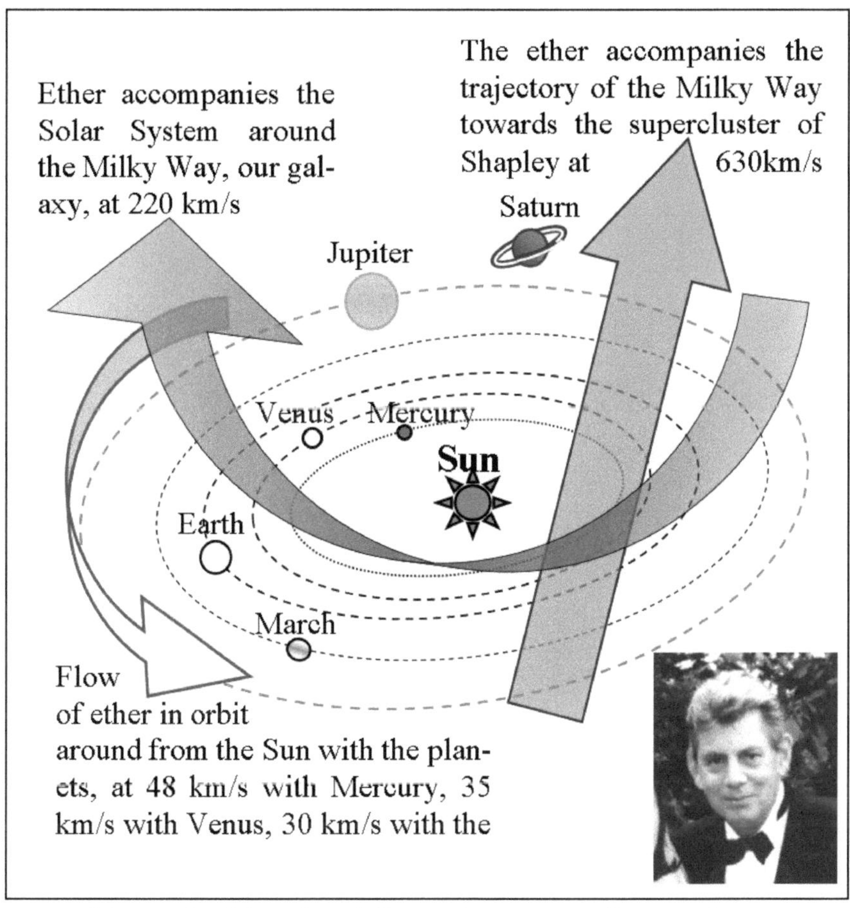

Figure 17. The ether follows the curves of space time

Above, we propose a simplified view of the ether's movements, which corresponds to its "immobility" in all reference frames in free fall.

This is all well and good, but it is only a hypothesis. A hypothesis must be verified. Who could refuse a hypothesis based on the reflections of Albert Einstein?

Experiments to perform

To check whether the ether accompanies the Earth, it suffices to use a reference frame moving near the latter and at a sufficiently high-speed relative to it to detect an ether wind. This reference frame must not be accompanied by a large mass capable of dragging the surrounding ether with it, which would cause the experiment to fail.

Let us take up Einstein's explanations on the relativity of lengths and times (§2 On the relativity of lengths and times) and replace the rigid rod AB with two satellites rotating in the plane of the ecliptic at low altitude and moving at speed v relative to the Earth. Let us place ourselves at the moment when the line connecting the two satellites is parallel to the trajectory of the Earth around the Sun, that is, at noon or midnight at the Sun in order to obtain the largest difference. If the ether accompanies the Earth around the Sun, we should detect a speed difference and, therefore, a difference between the time taken by electromagnetic waves to go from A to B and the time necessary to return from B to A.

If the satellites are separated by a sufficiently large distance, we can hope that the quantity of ether that they will carry with them on their geodesic will be limited to their immediate environment and that over the majority of the distance separating them, the ether will be motionless relative to the Earth.

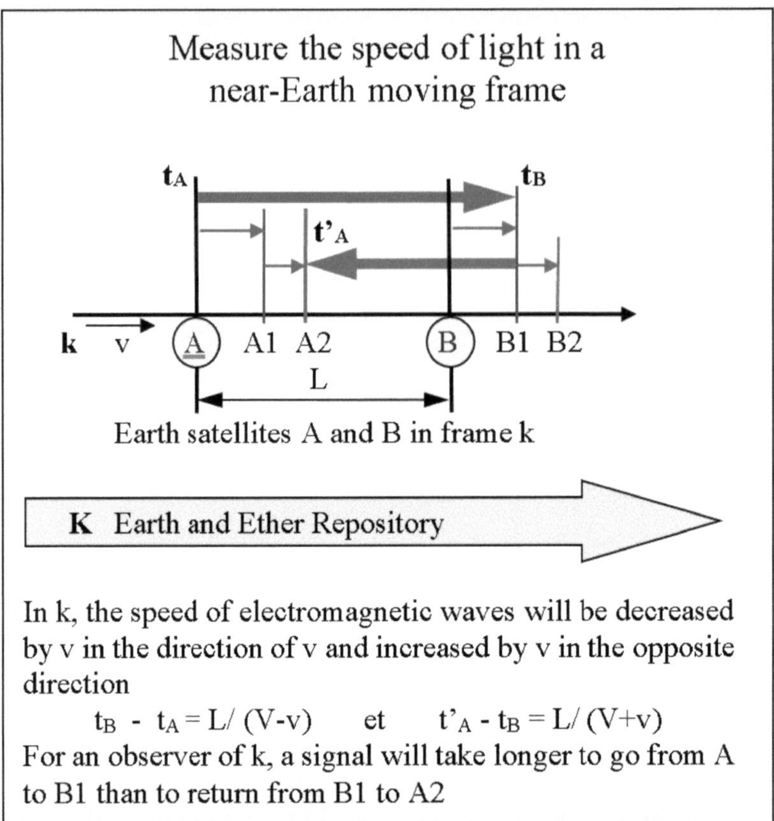

Figure 18. Measurement of outward and return time

In this diagram, the Earth is located in the reference frame K, which is deemed to be stationary. The reference frame k is that of the satellites; it moves with respect to K at the speed v without the ether accompanying it on its geodesic.

The speed of light is no longer a universal constant in reference frames associated with low masses and moving near large masses.

Einstein in his 1905 article explains that seen from k the speed of light is the same going out and coming back. This would be true if the ether were in the reference frame k

Satellites in low orbit

Two nanosatellites are orbiting at low altitude in the plane of the ecliptic at around 9 km/s, 1,000 km apart. When they are parallel to the trajectory of the Earth, measure the speed of the waves from A to B then from B to A.

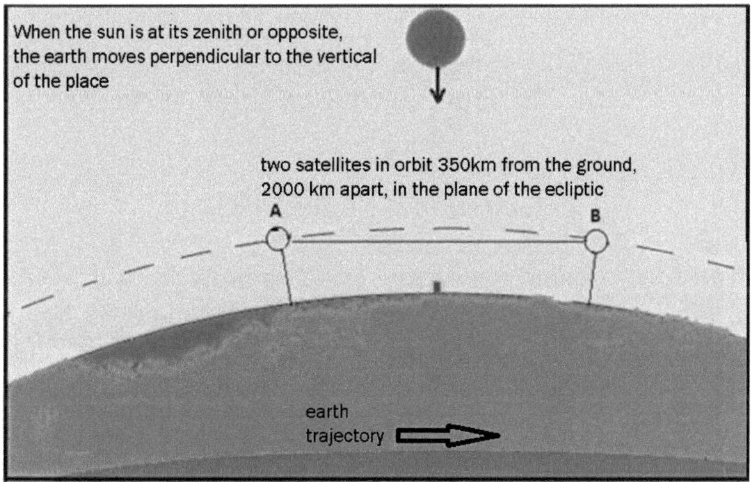

Figure 19. Two satellites in low orbit

Probe rotating in Earth's orbit in the opposite direction

Probe rotating in Earth's orbit in the opposite direction would be the most compelling experience. If the carrier medium of the electromagnetic waves accompanies the Earth at 30 km/s around the Sun, the probe would move at 60 km/s in relation to it. This would also be an opportunity to check the Unruh effect.

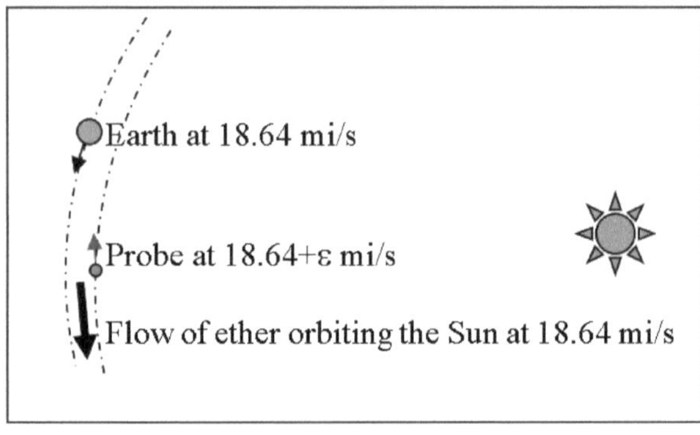

Figure 20. Probe rotating on Earth's orbit

International Space Station

The time it takes for a radio wave emitted from the International Space Station (ISS) to reach a ground station can be measured when the motion of the ISS is closest to the Earth's path and when it is perpendicular to this trajectory (noon or midnight at the Sun).

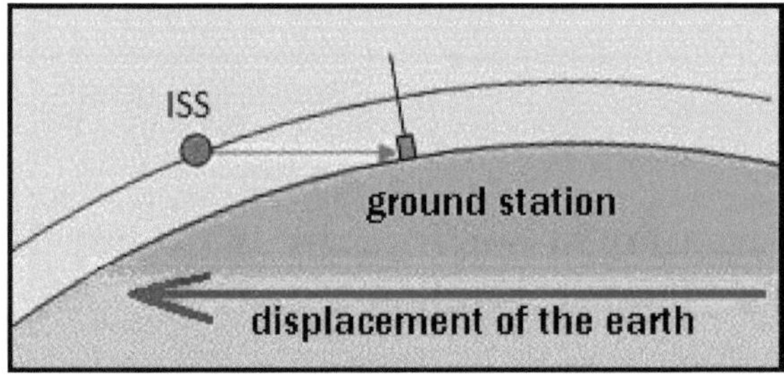

Figure 21. Link between the ISS and a ground station

Inclination of the ISS orbit relative to the equator: 51.65°, relative to the ecliptic: 28.39°, ISS horizon at 350 km altitude: 2,141 km

Of course, we saved the best for last by using a shuttle approaching the ISS around noon at the Sun.

The simplest experiment!

The simplest experiment is to take advantage of the arrival of a shuttle carrying astronauts at the ISS. When it is at a distance of about 1,000 km and the line connecting it to the ISS is parallel to the movement of the Earth, we measure the outward and return time of an electromagnetic signal from the shuttle to the ISS, and vice versa.

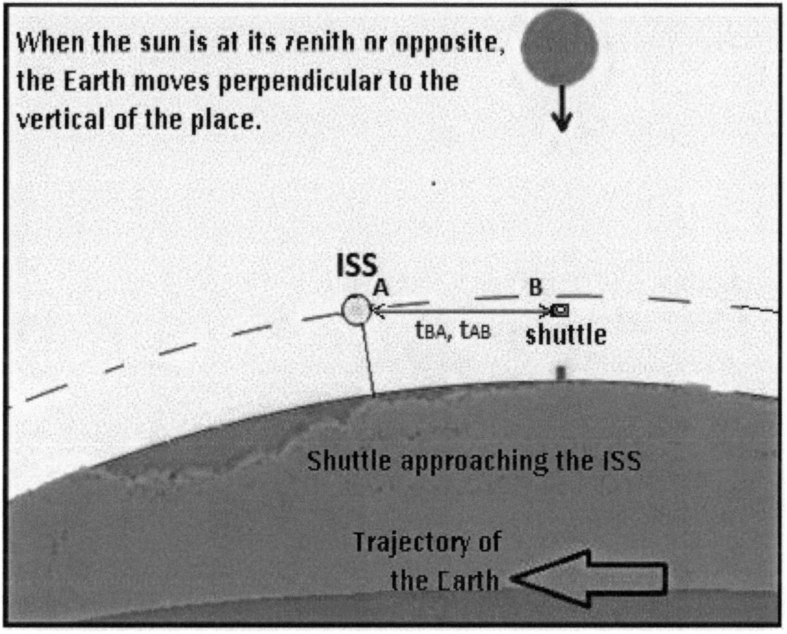

Figure 22. Link between the ISS and an approaching shuttle

Measurements at different positions in the orbit of the ISS make it possible to determine the minimum and maximum deviations. The estimate of the time deviations to be detected depends on the projection of the speed of the ISS on the tangent to the trajectory of the Earth around the Sun and the distance of the shuttle.

We will leave it to the specialists to do their best.

Epilogue

Albert Einstein was the first to realize that ether cannot be stationary. It must preserve the universality of the speed of light and move with all reference frames in the same way so as to respect the principle of relativity extended to electromagnetics. Our hypothesis solves the motion condition fixed by Albert Einstein but does not allow the speed of light to be maintained as a universal constant in all circumstances, but only in cases where measurements are made in reference frames of large masses and at proximity to them. Validation by experiments will be our judge as is always the case in the scientific field.

We suggest that the proposed experiment between the ISS and a shuttle, Dragon or Soyuz, which should be feasible without prohibitive cost, be the first to be organized. Failing that, we suggest a link with a ground station, if necessary with a team of radio amateurs from a school, who will have to furnish their equipment with high-precision clocks and who will have to establish a link between 11 a.m. and 1 p.m. in the Sun or 11 p.m. and 1 a.m.

Isaac Newton said he was inspired by watching an apple fall. We, in our adolescence, like Pierre Barouh, Françoise Hardy, and a few others, took pleasure at the edge of streams in making circles in the water. Allow the dreamers, failing to match the giants, to climb on their shoulders and gaze into the distance.

It may be that by dint of searching, they see by chance some wonders of the universe.

Picture table

Notes

1) The Immobility of the Earth in the Center of the World §8
https://www.originalsources.com/document.aspx?DocID=E847WGAPL4RHJNM

2) Dialogue Concerning the Two Chief World Systems—Ptolemaic and Copernican
https://www.britannica.com/topic/Dialogue-Concerning-the-Two-Chief-World-Systems-Ptolemaic-and-Copernican

3) https://en.wikipedia.org/wiki/Permittivity
https://en.wikipedia.org/wiki/Permittivity

4) Michelson-Morley experiment:
https://en.wikipedia.org/wiki/Michelson%E2%80%93Morley_experiment

5) Lorentz_transformation
https://en.wikipedia.org/wiki/Lorentz_transformation

6) Woldemar Voigt
https://en.wikipedia.org/wiki/Woldemar_Voigt

7) Henry Poincaré
https://en.wikipedia.org/wiki/Henri_Poincar%C3%A9
Henri Poincaré, mémoire de Palerme
https://www.lajauneetlarouge.com/24-septembre-1904-le-principe-de-relativite-dhenri-poincare-1854-1912/

8) Founding article of special relativity, published in German in 1905 in the journal Annalen der Physik, under the title: "Zur Elektrodynamik bewegter Körper" English translation: « On the electrodynamics of moving bodies» p. 1
https://www.fourmilab.ch/etexts/einstein/specrel/www/

9) On the electrodynamics of moving bodies p.1 at the bottom

10) Photoelectric effect
https://en.wikipedia.org/wiki/Photoelectric_effect

11) On the electrodynamics of moving bodies Kinematical part § 1. Definition of Simultaneity

12) On the electrodynamics of moving bodies Kinematical part §
    1. Definition of simultaneity at the end

13) On the electrodynamics of moving bodies § 2. On the relativity of lengths and times.

14) On the electrodynamics of moving bodies § 3. Theory of the transformation of co-ordinates

15) Twin paradox https://en.wikipedia.org/wiki/Twin paradox

16) Einstein "the happiest idea of my life"
    https://en.wikipedia.org/wiki/Local_reference_frame

17) The principle of equivalence between acceleration and gravitation https://en.wikipedia.org/wiki/Equivalence
    princile#:~:text=The%20equivalence%20principle%20was%
    20properly,to%20the%20acceleration%20of%20an

18) Patrick Girard. Histoire de la Relativité Générale d'Einstein:
    Développement Conceptuel de la Théorie. Sciences de
    l'Homme et Société. University of Wisconsin-Madison
    USA, 1981. Français tel-02321688 https://hal.archives-
    ouvertes.fr/tel-02321688/document

19) https://www.google.fr/search?tbm=bks&hl=fr&q=In+G%C3
    %B6ttingen+I+had+the+great+joy+to+see+everything+unde
    rstood+and+even+in+detail.+I+am+very+enthusiastic+about
    +Hilbert.+A+great+man

20) Gravitational wave
    https://en.wikipedia.org/wiki/Gravitational wave

21) Lorentz ether theory
    https://www.chemeurope.com/en/encyclopedia/Lorentz_ethe
    r_theory.html

22) Einstein admits the existence of ether
    https://www.academia.edu/986494/Einsteins_Ether_F_Why_
    did_Einstein_Come_Back_to_the_Ether (Ludwik Kostro,
    Einstein and the ether, Montreal, Apeiron, 2000 (ISBN 0-
    9683689-4-8)

23) Einstein's New Ether Springer note 30 de) A. Einstein, Äther
    und Relativitätstheorie: Exposé fait le 5 mai 1920 à l'Univer-
    sité de Leyde, Berlin, 1920

http://www.mathem.pub.ro/proc/bsgp-10/K10-KOSTRO.PDF
24) Einstein, Äther und Relativitätstheorie https://zionism-isra-el.com/Albert_Einstein/Albert_Einstein_Ether_Relativity.htm
25) Thibault Damour: Space is a block of jelly https://www.ihes.fr/~damour/Conferences/OG_TROUS_NOIRS_Luxembourg2016.pdf
26) Unruh effect https://en.wikipedia.org/wiki/Unruh_effect
27) Casimir effect *https://physicsworld.com/a/the-casimir-effect-a-force-from-nothing/*

Table des matières